Living Things Depend on One Another

Lesson 1
How Do Plants and Animals Interact? ... 2

Lesson 2
What Are Food Chains? ... 10

Lesson 3
What Are Food Webs? ... 18

Harcourt
SCHOOL PUBLISHERS

Orlando   Austin   New York   San Diego   Toronto   London

Visit *The Learning Site!*
www.harcourtschool.com

Lesson 1

How Do Plants and Animals Interact?

VOCABULARY
producer
consumer
decomposer
herbivore
carnivore
omnivore

A **producer** is a living thing that makes its own food. Plants are producers.

A **consumer** is a living thing that gets food by eating plants or animals. Animals are consumers.

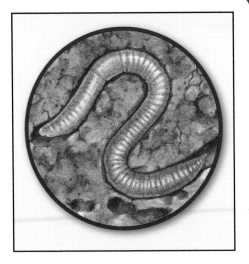

A **decomposer** is a living thing that breaks down dead things for food. An earthworm is a decomposer.

A **herbivore** is an animal that eats only plants. A caterpillar is a herbivore.

A **carnivore** is an animal that eats only other animals. A wolf is a carnivore.

An **omnivore** is an animal that eats both plants and animals. A chicken is an omnivore.

READING FOCUS SKILL
COMPARE AND CONTRAST

When you **compare and contrast**, you tell how things are alike and different.

Look for ways to **compare and contrast** how different animals get energy.

Making and Getting Food

All living things need food to get energy. Plants make their own food. They use sunlight, air, and water to make it. Plants store the food they do not use. This food is often full of energy.

Animals cannot make their own food. They get food by eating plants or other animals. The energy in these foods goes into their bodies.

 Tell how how plants and animals get food in different ways.

◀ Plants get energy from the sun. Rabbits get energy by eating plants.

Producers and Consumers

Plants are producers. A **producer** is a living thing that makes its own food. Producers use this food to grow.

Animals are consumers. A **consumer** gets energy by eating other living things. Consumers eat plants or other animals. A consumer cannot make its own food.

 Tell how producers and consumers are alike and different.

A sunflower is a producer. A bird is a consumer.

Decomposers

Some living things are decomposers. A **decomposer** breaks down dead things for food. Earthworms and mushrooms are decomposers. Most decomposers are very small. You can see them only with a microscope.

 Tell how decomposers and producers are alike and different.

▼ Earthworm

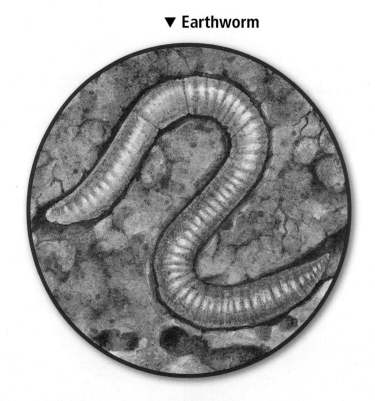

Herbivores

There are three kinds of consumers. They are called herbivores, carnivores, and omnivores.

A **herbivore** is a consumer that eats only plants. Herbivores can be small, like bees. They can also be large, like horses.

Each herbivore has body parts that help it eat plants. Cows and horses have flat teeth. Their teeth help them grind up grass.

 Tell how a herbivore is different from a producer.

Horses eat grass. ▼

This Galápagos (guh•LAH•puh•gohs) tortoise eats only plants. ▼

▲ Wolves eat only meat.

Carnivores

A **carnivore** is a consumer. It gets food by eating other animals.

Carnivores have body parts that help them hunt and eat. Wolves have sharp teeth to eat meat. Some birds have sharp claws and strong beaks. These body parts help them catch and eat food.

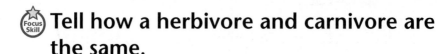 Tell how a herbivore and carnivore are the same.

Omnivores

An **omnivore** is a consumer that eats both plants and animals. Bears are omnivores.

Most omnivores have teeth that help them eat plants and animals. They have sharp teeth to help them eat meat. They also have flat teeth to help them grind up plants.

How are herbivores, carnivores, and omnivores different?

Raccoons are omnivores. ▶

Review

Complete these compare and contrast statements.

1. All _____ need food.

2. Unlike consumers, _____ can make their own food.

3. Both carnivores and _____ eat other animals.

Lesson 2

VOCABULARY
food chain
energy pyramid
predator
prey

What Are Food Chains?

Grass   Grasshopper   Lizard

A **food chain** shows how energy moves from one living thing to another. It does this by showing what animals eat.

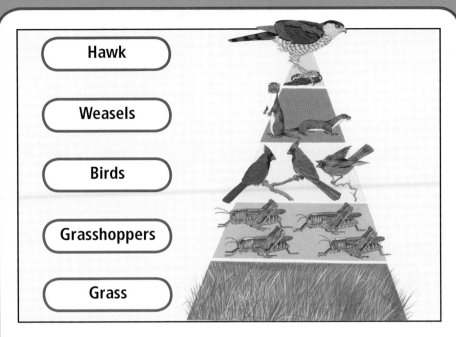

An **energy pyramid** shows how energy is used in a food chain.

A **predator** is an animal that hunts another animal for food. The animal that is hunted is called **prey**. This cheetah is the predator and the gazelle is the prey.

READING FOCUS SKILL

SEQUENCE

A **sequence** is the order in which things happen.

Look for the **sequence** of how energy passes through a food chain.

Food Chains

A **food chain** shows the path of food from one living thing to another. Use the arrows below to see what each animal eats.

▲ Grass is a producer. It uses sunlight to make its own food.

▲ The grasshopper eats the grass. It gets energy stored in the grass.

A food chain starts with a producer. Grass is the producer in this food chain. It uses energy from the sun to make its food. This energy is passed on in a food chain.

Some food chains are long. Some are short. People are at the end of many food chains.

What happens to the energy in grass after the grasshopper eats it?

▲ This lizard eats the grasshopper. It gets energy from the grasshopper.

▲ This owl eats the lizard. It gets energy from the lizard.

Energy in a Food Chain

Energy passes from one living thing to another in a food chain. When animals eat plants, they get the energy stored in plants. Animals use some of the energy to live and grow. The rest is stored in their bodies. Other animals get that energy when they eat them.

 What happens to the energy in plants?

This bobcat must eat many smaller animals to get the energy it needs. ▶

Energy Pyramid

An **energy pyramid** shows how energy is used in a food chain. It shows that very little energy is passed up from one level to the next. That is why each level has fewer living things than the one below it. Producers, or plants, are at the bottom of an energy pyramid. They are the biggest group.

Tell what happens to the energy from the birds.

Predator and Prey

A **predator** is an animal that hunts another animal for food. A lynx is a predator. An animal hunted for food is called **prey**. A rabbit is prey for a lynx.

Some animals are predators and prey. Birds that eat insects are predators. But these birds may be prey for a hawk.

 Tell what happens to prey.

Puffins are predators that eat many fish. ▶

Review

 Complete these sequence sentences telling about a food chain.

1. A food chain always begins with a _____.

2. A _____ becomes part of a food chain after a plant or animal dies.

3. Living things get _____ stored in the plants and animals that come before them in a food chain.

Lesson 3
What Are Food Webs?

VOCABULARY
food web

A **food web** is made up of food chains that overlap.

READING FOCUS SKILL
MAIN IDEA AND DETAILS

A **main idea** is what the text is mostly about.
Details tell more about the **main idea**.
Look for **details** about food webs.

A Marsh Food Web

Food Web

An animal may eat many things. This means it is part of several food chains. A **food web** shows how many food chains overlap.

Look at this food web. Use the arrows to find the foods each animal eats.

⭐ **Tell what a food web is.**

▲ Wildebeests live in big herds to stay safe.

Ways Animals Defend Themselves

Animals defend themselves in many ways. An animal's color or shape may help it hide. Its teeth, sting, or bad smell can protect it. It may run fast. It may play dead, too.

Focus Skill Tell three ways animals defend themselves.

▼ A snake may bite.

▼ Oppossums play dead.

Changes in Food Webs

Many things can change a food web. A change to one part can change other parts.

Adding new plants can change a food web. New plants may crowd out other plants. Adding new animals can change a food web, too. New animals may eat other animals or plants.

 What can change a food web?

Review

Complete this main idea statement.

1. A _____ is made up of food chains that overlap.

Complete these detail sentences.

2. Animals have many ways to _____ themselves from predators.

3. A food web can change if _____ or animals are added to it.

GLOSSARY

carnivore (KAHR•nuh•vawr) an animal that eats other animals

consumer (kuhn•SOOM•er) a living thing that gets its energy by eating other living things as food

decomposer (dee•kuhm•POHZ•er) a living thing that breaks down dead organisms for food

energy pyramid (EN•er•jee PIR•uh•mid) a diagram that shows how energy gets used in a food chain

food chain (FOOD CHAYN) the path of food from one living thing to another

food web (FOOD WEB) food chains that overlap

herbivore (HER•buh•vawr) an animal that only eats plants

omnivore (AHM•nih•vawr) a consumer that eats both plants and animals

predator (PRED•uh•ter) an animal that hunts another animal for food

prey (PRAY) an animal that is hunted by another animal, a predator

producer (pruh•DOOS•er) a living thing that makes its own food